Get set... GO!

Puff and Blow

Sally Hewitt

Photography by Peter Millard

Contents

CHILDRENS PRESS®
CHICAGO

Introduction

Sound waves are made by
making air vibrate.
This means the air moves back and
forth very, very fast.
You cannot see sound waves,
but you can hear them.

Players of wind instruments make air
vibrate by blowing.
All the instruments in the picture are
wind instruments.
They have hollow tubes.
When you blow into these hollow tubes,
the air inside them vibrates and plays a note.

Get ready to make some instruments
to puff and blow.

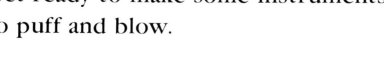

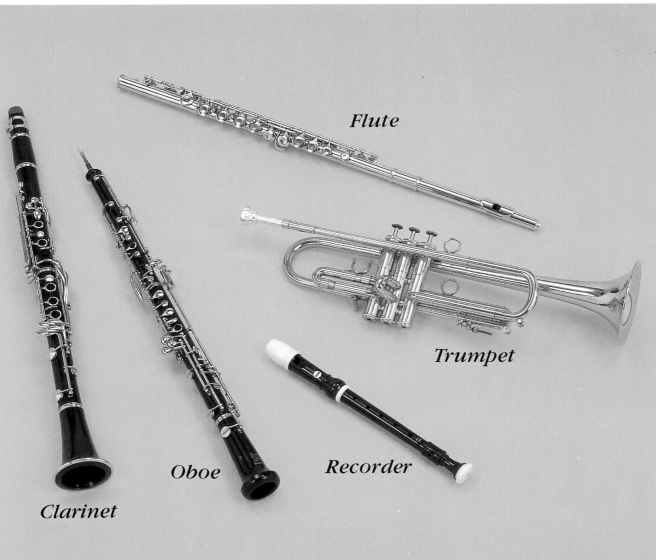

Flute

Trumpet

Oboe

Recorder

Clarinet

Comb and paper

Get ready

✔ Comb

✔ Tissue paper, tracing paper, or waxed paper

. . . Get Set

Fold the paper over the comb.
Hold them both against your mouth.

 *Go!*

Sing a tune loudly.
Feel the paper tickle your lips.
Your breath makes the paper vibrate.
The moving paper changes
the sound of your voice.
Listen to it buzz and hum!

Singing balloon

Get ready

✔ Balloon

. . . Get Set

Blow up the balloon.

≡❀≡❀≡❀ *Go!*

Pinch both sides of the balloon's neck.
Let the air out gradually.
As it escapes, the air vibrates
and makes a sound.
You can make different sounds.
Let the air out of a small opening
to make a high squeal.
Let the air out of a bigger opening
for a lower wail.

Bottles and tubes

Get ready

✔ Bottles of different sizes
✔ Tubes of different lengths

. . . Get Set

Group the bottles by size (big or small).
Group the tubes by length (long or short).

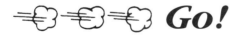 *Go!*

Blow gently across the tops of
the bottles and tubes.
The air inside them vibrates
and plays a note.
Long tubes and big bottles
make low notes.
Shorter tubes and smaller bottles
make higher notes.

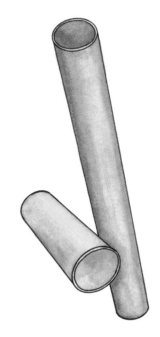

Bottle pipes

Get ready

✔ Several bottles of the same size
✔ Water

. . . Get Set

Blow across the top of each bottle.
They all make the same hooting sound.
Pour different amounts of water into the bottles.
Now there is a different amount of air
inside each bottle.

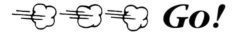 **Go!**

Blow across them again.
The bottle with the most air
plays the lowest note.
The bottle with the least air
plays the highest note.

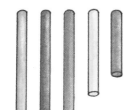

Panpipes

Get ready

✔ Large drinking straws   ✔ Masking tape
✔ Safety scissors

. . . Get Set

Blow across the top of the straws.
Listen to the air vibrate inside them
and make a sound.

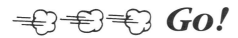 *Go!*

Cut the straws in different lengths.
Blow across each one and listen to its
sound.
Short straws play high notes.
Longer straws play lower notes.
Put the straws in order of size.
Tape them together to make panpipes.

Wailer

Get ready

✔ A long copper pipe   ✔ Wooden spoon

. . . Get Set

Make sure the handle of the spoon
fits inside the pipe.

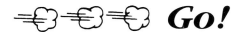 **Go!**

Blow across the top of the pipe.
As you blow, push the handle
of the spoon up and down
inside the pipe.
This changes the amount of air
inside the pipe.
Listen to the strange wailing sound.

Loud and soft

Get ready

✔ Facial tissue   ✔ Cardboard tube

. . . Get Set

Hold the tissue in front of your mouth.

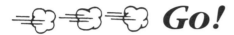 *Go!*

Sing the sound "ooo" very loudly.
Watch the tissue move.
Now sing the sound "ooo" very softly.
The tissue hardly moves at all.
You use more breath to sing loudly
than to sing softly.
Stuff the tissue into the end of the tube.
Sing loudly into the tube.
The tissue traps the air inside the tube
and muffles the sound.

Reed

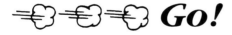

Get ready

✔ Long blade of grass
(or thin piece of paper)

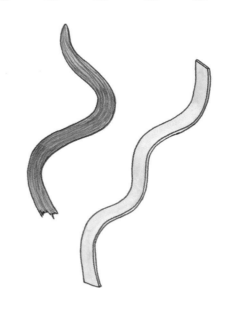

. . . Get Set

Wash the blade of grass
and dry it carefully.

⊰⊱⊰⊱⊰⊱ *Go!*

Hold the blade of grass firmly
between your thumbs.
Cup your hands.
Press your lips over
the space between your thumbs.
Blow hard.
The grass vibrates and makes
a strange squeaking noise.

18

Buzz and hum

Get ready

✔ Sing

. . . Get Set

Feel your voice coming from
your chest and head.

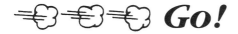 **Go!**

Put your hand on your chest.
Make a buzzing sound like a bee.
Feel the vibrations.
The sound comes through your mouth.
Now press your lips together
and hum like a top.
Feel the vibrations in your head.
This sound comes through your nose.

Tongue and lips

Get ready

✔ Waggle your tongue ✔ Move your lips

. . . Get Set

Take a deep breath and sing a long note.

Go!

Change the shape of your lips to make
"ooo," "aaa," and "eee" sounds.
Waggle your tongue to make
"lubba-lubba" sounds.
Flap your lips with your finger.
How many different sounds
can you make by moving
your tongue and lips?

Index

Acknowledgments: The author and publisher would like to thank the pupils of Kenmont Primary School, London, for their participation in the photographs of this book.

Editor: Pippa Pollard
Design: Ruth Levy
Cover design: Mike Davis
Artwork: Ruth Levy

Library of Congress Cataloging-in-Publication Data
Hewitt, Sally.
    Puff and blow / by Sally Hewitt.
    p.   cm. — (Get set— go!)
    Includes index.
    ISBN 0-516-07993-X
    1. Sound-waves—Juvenile literature. 2. Sound-waves—Experiments—Juvenile literature. 3. Music—Acoustics and physics—Juvenile literature. 4. Music—Acoustics and physics—Experiments—Juvenile literature. [1. Sound—Experiments. 2. Music—Experiments.] I. Title. II. Series.
QC243.2.H49   1994
788'.19—dc20   94-16910
                              CIP
                              AC

1994 Childrens Press® Edition
© 1993 Watts Books, London, New York, Sydney
All rights reserved. Printed in the United States of America.
Published simultaneously in Canada.
1 2 3 4 5 6 7 8 9 0 R 03 02 01 00 99 98 97 96 95 94